ON THE NATURAL HISTORY OF

THE ARU ISLANDS

(1857)

BY

ALFRED RUSSEL WALLACE

British Library Cataloguing-in-Publication Data
A catalogue record for this book is available from the
British Library

Alfred Russel Wallace

Alfred Russel Wallace was born on 8th January 1823 in the village of Llanbadoc, in Monmouthshire, Wales.

At the age of five, Wallace's family moved to Hertford where he later enrolled at Hertford Grammar School. He was educated there until financial difficulties forced his family to withdraw him in 1836. He then boarded with his older brother John before becoming an apprentice to his eldest brother, William, a surveyor. He worked for William for six years until the business declined due to difficult economic conditions.

After a brief period of unemployment, he was hired as a master at the Collegiate School in Leicester to teach drawing, map-making, and surveying. During this time he met the entomologist Henry Bates who inspired Wallace to begin collecting insects. He and bates continued exchanging letters after Wallace left teaching to pursue his surveying career. They corresponded on prominent works of the time such as Charles Darwin's *The Voyage of the Beagle* (1839) and Robert Chamber's *Vestiges of the Natural History of Creation* (1844).

Wallace was inspired by the travelling naturalists of the day and decided to begin his exploration career collecting specimens in the Amazon rainforest. He explored the Rio Negra for four years, making notes on the peoples and languages he encountered as well as the geography, flora, and fauna. On his return voyage his ship, Helen, caught fire and he and the crew were stranded for ten days before being picked up by the Jordeson, a brig travelling from Cuba to London. All of his specimens aboard Helen had been lost.

After a brief stay in England he embarked on a journey to the Malay Archipelago (now Singapore, Malaysia, and Indonesia). During this eight year period he collected more than 126,000 specimens, several thousand of which represented new species to science. While travelling, Wallace refined his thoughts about evolution and in 1858 he outlined his theory of natural selection in an article he sent to Charles Darwin. This was published in the same year along with Darwin's own theory. Wallace eventually published an account of his travels *The Malay Archipelago* in 1869, and it became one of the most popular books of scientific exploration in the 19th century.

Upon his return to England, in 1862, Wallace became a staunch defender of Darwin's landmark work *On the Origin of Species* (1859). He wrote responses to those critical of the theory of natural selection, including 'Remarks on the Rev. S. Haughton's Paper on the Bee's Cell, And on the Origin of Species' (1863) and 'Creation by Law' (1867). The former of these was particularly pleasing to Darwin. Wallace also published important papers such as 'The Origin of Human Races and the Antiquity of Man Deduced from the Theory of 'Natural Selection" (1864) and books, including the much cited *Darwinism* (1889).

Wallace made a huge contribution to the natural sciences and he will continue to be remembered as one of the key figures in the development of evolutionary theory.

Wallace died on 7th November 1913 at the age of 90. He is buried in a small cemetery at Broadstone, Dorset, England.

ON THE NATURAL HISTORY OF THE ARU ISLANDS
(1857)

In December 1856, I left Macassar in one of the trading prows which make an annual voyage to these islands. On January 1st, 1857, we arrived at the Ké Islands. Here we remained six days, while the natives, who are clever boat-builders, finished two small vessels our captain purchased for the Aru trade. During this time I made daily excursions in the forests, collecting birds and insects; but the weather was showery, and the coralline-limestone rocks, which everywhere protrude through the thin soil, are weather-worn into such sharp-edged, honey-combed, irregular surfaces, as to make any distant excursions almost impossible. The great Fruit Pigeon of the Moluccas (*Carpophaga ænea*) was abundant, its loud, hoarse cooings constantly resounding through the forest. Crimson Lories of two or three species were also plentiful, but were so wary that we could not obtain any. All other birds were scarce, and I only obtained thirteen species in all, many of which will, however, I think, prove new, viz.:--

Megapodius, sp., same species at Aru.

Carpophaga ænea.

7

Ptilonopus, n. sp.

Macropygia, sp., same at Aru.

Dicrurus, sp.

Tropidorhynchus, n. sp.?

Cinnyris, n. sp.

Zosterops citrinella, Müll.

Rhipidura, two species.

Muscicapidæ aliæ, two species.

Psittacus (*Geoffroyus?*), sp.

Among the birds offered for sale, *Eclectus Linnæi* and *Psittacodis magnus* were the most abundant. Of Mammalia I saw none, and could only learn that a wild pig and a species of *Cuscus* inhabited the island. The only reptiles I saw were lizards of two or three kinds, one of which, a very long and slender species, with a finely-pointed tail of a most brilliant blue, swarmed everywhere on the low herbage, gliding among leaves and twigs in the most rapid and elegant manner. Of insects I made a nice little collection, the natives bringing me several very fine Coleoptera. A considerable proportion appear to be quite new; those known being a mixture of New Guinea and Molucca species. It would occupy too much space to enter into any details on this extensive class; I shall therefore give only the results of my six days' work, as follows:--

Coleoptera	70 species.
Lepidoptera	50 "
Diptera	19 "
Hymenoptera	24 "
Hemiptera and others	31 "
	194 species of insects

It was here I first made acquaintance with the Papuan race in their native country, and it was with the greatest interest I studied their physical and moral peculiarities, and noted the very striking differences that exist between them and the Malays, not only in outward features, but in their character and habits.

A day and a half's sail brought us to the trading settlement of Dobbo, situated on a sandy spit running out in a northerly direction from the island of Wamma, which here approaches to within a mile of the great island of Aru. Having obtained the use of one of the palm-thatched sheds here dignified with the name of houses, arranged my boxes and table, and put up a drying-shelf indoors and out, protected by water-insulation from the attacks of ants, I was ready to commence my exploration of the unknown fauna of Aru. I had brought with me two boys, whose sole business was to shoot and skin birds, while I attended entirely to insects, and to the observation and registry of the habits of the birds and animals I met with in my walks in the forest. The first fortnight was very unpropitious, violent gusts of wind and driving rain allowing us to do very little out of doors, and making the drying of the little we obtained a matter of great difficulty. It soon became apparent that in this small island there was a very limited number of birds, and I determined to go as soon as possible to the large island; but that was not an easy matter, and I now found that I should have brought from Macassar three men accustomed to the islands, and who could take me wherever I wanted to go. As it was, I had to get natives, and there was, as usual, all sorts of delay, and then there was an alarm of pirates, and unfortunately it was not a false alarm. A fleet of the celebrated Ilanun pirates, from the island of Maguidanao, had really arrived; they attacked a small vessel not far from Dobbo, which, escaping from them with one man wounded, brought the news. Then came messengers from one of the northern islands, telling how they had been attacked, and many taken prisoners, and the rest of the population had all fled to the mainland. Now for some time there was no more hope of my getting boats and men. Guards were set in Dobbo, and prows were got ready to go after the pirates. A few days more, and the crew of one of our captain's small vessels which had gone trading among the islands, returned stripped of everything. They had got on shore, while the pirates plundered the prow, taking everything, even to the men's boxes and clothes. They reported that the pirates were all at the east side of the islands, where the merchants send their small vessels to buy pearl-shell and tripang, and there was no danger of their returning again to this side, where they had more to fear and less to get. Now, too, I received a letter I had been expecting from the Governor of Amboyna, with orders to the Aru chief to give me assistance; and, after two months' residence in Dobbo, I succeeded in getting a boat and two natives, and set off for the great island of Aru.

I visited several localities, and at length, finding a good one near the centre of the island, I stayed there six weeks, and got, on the whole, a very fine collection of birds. Returning to Dobbo, I intended to make another short excursion; but lameness, produced by the constant irritation of insect-bites on my legs, kept me in the house for several weeks, and the east wind became so strong, and the weather so wet and boisterous, as to render travelling by sea in a small boat out of the question. A little later, one of my bird-skinners left me, and the other was laid up with intermittent fever, so I was compelled to make the best of it, and get what I could in the small island till the commencement of July, when we returned direct to Macassar.

Having thus given an outline of my journey, I shall proceed to give some account of the ornithology and general natural history of the Aru Islands, and a summary of the collections I have made there. The very first bird likely to attract one's attention at Dobbo is a most beautiful brush-tongued parroquet, closely allied to *Trichoglossus cyanogrammus*, Wagl. It frequents in flocks the Casuarina-trees which line the beach, and its crimson under wings and orange breast make it a most conspicuous and brilliant object. Its twittering whistle may be heard almost constantly in the vicinity of the trees it frequents. Almost the only other birds which approach the village are a swallow (*Hirundo nigricans*, Vieill.), found also in New Guinea and Australia, and an *Artamus*, probably *A. papuensis*, Temm., which perches occasionally on the house-tops, or on dead trees in the neighbourhood. A little black-and-white wagtail flycatcher (*Rhipidura*, sp.) may also often be seen among bushes, and on the sea-beach, chirping musically, and waving laterally its expanded tail whenever it alights.

In the forest which everywhere covers the islands, sombre and lofty as on the banks of the Amazon, a different set of birds is met with, the two most abundant being both New Guinea species, *Cracticus varius*, Gm., sp., and *Phonygama viridis*, L. The former has a loud and very varied note; sometimes a fine musical whistle; at others (principally when alarmed), a harsh, toad-like croak. It is very active, flying about from tree to tree and from bush to bush, seeking after insects, or feeding on small fruits. It is a long time before one can recognize its various cries for those of one and the same bird. The *Phonygama* is a very powerful and active bird; its legs are particularly strong, and it clings suspended to the smaller branches, while devouring the fruits on which alone it appears to feed. Its affinities seem to be with the Paradiseas rather than with the Garrulidæ. Next to these, two species of *Dacelo* are most frequently met with, and their loud monotonous cry, very much resembling the bark of a dog, most frequently heard. A large crow, with a fine sky-blue iris, and hoarse cawing cry, is also not uncommon; and now I have mentioned all the birds, except parrots and pigeons, that are common enough to be at all characteristic of the forest near Dobbo. For noise, however, the Psittacidæ surpass all others, and the Yellow-crested Cockatoo (*Plyctolophus galerita*) is an absolute nuisance. Instead of flying away when alarmed, as other birds do, it circuits round and round from one tree to another, keeping up such a grating, creaking, tympanum-splitting scream, as to oblige one to retire as soon as possible to a distance. Far more agreeable is the low cooing of the pigeons, several fine species of which are not uncommon. *Carpophaga pinon*, Q. & G., is plentiful, and another, which seems to be *C. Zoë*, Less., rather scarce; while *C. alba*, L., is common everywhere. Of the smaller and more beautiful species there are also three, *Ptilonopus perlatus*, Temm., *P. pulchellus*, Temm., and *P. purpuratus*, Lath. These birds are all very difficult to obtain in good condition, because their feathers fall so readily; but they are always acceptable, as their flesh (especially that of the smaller species) is perhaps equal in delicacy and flavour to that of any birds whatever.

In one or two excursions which I made to the mainland,

immediately opposite Dobbo, I obtained the two beautiful flycatchers, *Arses telescophthalma*, Garn. & Less., and *A. chrysomela*, G. & L., as well as some species of *Ptilotis* and other small birds new to me. It was not, however, until I was regularly established in the central forests of the large island that I obtained a true insight into the ornithological fauna of Aru. Then a host of new species burst upon me, revealing the richness of the country, and its intimate connexion with New Guinea. *Paradisea apoda*, L., *P. regia*, L., *Microglossus aterrimus*, Wagl., *Brachyurus Macklotti*, Temm., *B. novæ-guineæ*, Schlegel, *Tanysiptera*, sp., *Eurystomus gularis*, Vieill., *Carpophaga*, n. s., with several small flycatchers, thrushes and shrikes, and that most magnificent of the swallow-tribe, *Macropteryx mystaceus*, Less., were what I now obtained,-- almost all New Guinea species, or new, and none of them found on the smaller islands. Of the beautiful little "King Bird of Paradise," I obtained several specimens in perfect plumage and excellent condition. It feeds, I believe, entirely on fruit, frequenting lofty trees in the deep forest, where it is very active, flying from branch to branch, shaking its wings, and expanding its beautiful fan-shaped breast-plumes.[1] When quite at rest, or feeding, these plumes are closed and concealed beneath the wing. Of the "Great Bird of Paradise," I have recorded my observations in a separate paper. The Black Cockatoo is a very curious bird, of most disproportionate form and dimensions. Its huge head certainly weighs as much as its whole body. The legs are very long and slender for the tribe, while its wings are large and powerful. Its cry is a shrill whistle, very different from that of most other cockatoos. The bill of the male is larger, and the apex more produced, than in the female; but the crest-plumes are equally long in both. The *Tanysiptera* is a Kinghunter, feeding on insects, worms, &c., which it picks up from the ground in the damp forest. Its coral-red bill is always dirty from this cause, and sometimes so incrusted with mud that the bird seems to have been actually digging for its food. The *Syma torotoro*, Less., also occurs, but much more rarely, and seems to have very similar habits. Two species of *Megapodius* are plentiful, and the immense mounds of earth and leaves formed by them are scattered all over the forest. These mounds are generally from 5 to 8 feet high, and from 15 to 30 feet in diameter. But the giant of the Aru forests is the Cassowary (*Casuarius galeatus*, Vieill.); it is by no means uncommon, and the young are brought in numbers to Dobbo, where they soon become tame, running about the streets, and picking up all sorts of refuse food. When very young, they are striped with broad lines of rich brown and pale buff. This gradually fades into a dull pale brown, and in the old bird changes to black. They *sit* down to rest on their tibæ, and lie down on their breast to sleep; they are very frolicsome, having mock fights, rolling on their backs, and leaping in a most ridiculous manner with all the antics of a kitten. The same species is said to be found in Ceram, and also in the small island of Goram, as well as in New Guinea. The following list shows the number of species in each of the principal tribes and families which I have observed in Aru:--

Grallæ and Natatores	12 One duck, near *Anas radjah*, Less.
Gallinaceæ	15 Twelve pigeons, *Alecthelia Urvillei*, Less.
Accipitres (Falconidæ)	4
Psittaci	10
Paradiseidæ	2 New Guinea species.
Cinnyridæ	5 Three New Guinea species.
Meliphagidæ	9 Six *Ptilotis*.
Sturnidæ	2 Both New Guinea species.
Corvidæ	1
Garrulidæ	1
Laniidæ	3 Two New Guinea species.
Turdidæ	6
Ptiddæ	2 Both New Guinea species.
Maluridæ	2
Oriolidæ	2
Artamidæ	1 A New Guinea species.
Muscicapidæ	13 Four or five New Guinea species.
Edoliidæ	6
Coraciadæ	1 *Eurystomus*, same at Macassar and Lombock.
Hirundinidæ	3 One New Guinea, one Australian.
Caprimulgidæ	2 *Podargus* and *Caprimulgus*.
Alcedinidæ	11 Four or five New Guinea species.
Cuculidæ	3 *Centropus* and *Chrysococcyx*.
Total species	116

From this list, and the preceding observations, it will be seen that many Australian genera and some species occur in Aru; while, considering the very small number of species known from New Guinea, and the necessarily very imperfect exploration of Aru in such a short time, the number of identical species is very remarkable. I believe that nearly one-half of the hitherto-described species of passerine birds from New Guinea will be found in my Aru collections, a proportion which we could only expect if all the species of the latter country inhabit also the former. Such an identity occurs, I believe, in no other countries separated by so wide an interval of sea, for the average distance of the coast of Aru from that of New Guinea is at least 150 miles, and the points of nearest approach upwards of 100. Ceylon is nearer to India; Van Diemen's Land is not farther from Australia, nor Sardinia from Italy; yet all these countries present differences more or less marked in their faunas; they possess each their peculiar species, and sometimes even peculiar genera. Almost the only islands possessing a rich fauna, but identical with that of the adjacent continent, are Great Britain and Sicily, and that circumstance is held to prove that they have been once a portion of such continents, and geological evidence shows that the separation had taken place at no distant period. We must, therefore, suppose Aru to have once formed a part of New Guinea, in order to account for its peculiar fauna, and this view is supported by the physical geography of the islands; for, while the fathomless Molucca sea extends to within a few miles of them on the west, the whole space eastward to New Guinea, and southward to Australia, is occupied by a bank of soundings at a uniform depth of about 30 or 40 fathoms. But there is another circumstance still more strongly proving this connexion: the great island of Aru, 80 miles in length from north to south, is traversed by three winding channels of such uniform width and depth, though passing through an irregular, undulating, rocky country, that they seem portions of true rivers, though now occupied by salt water, and open at each end to the entrance of the tides. This phænomenon is unique, and we can account for their formation in no other way than by supposing them to have been once true rivers, having their source in the mountains of New Guinea, and reduced to their present condition by the subsidence of the intervening land.

This view of the origin of the Aru fauna is further confirmed by considering what it is not, as well as what it is; its deficiencies teach as much as what it possesses. There are certain families of birds highly characteristic of the Indian Archipelago in its western and better-known portion. In the Peninsula of Malacca, Sumatra, Java, Borneo and the Philippine Islands, the following families are abundant in species and in individuals. They are everywhere *common birds*. They are the *Buceridæ, Picidæ, Bucconidæ, Trogonidæ,*

Meropidæ, and *Eurylaimidæ*; but not one species of all these families is found in Aru, nor, with two doubtful exceptions, in New Guinea. The whole are also absent from Australia. To complete our view of the subject, it is necessary also to consider the Mammalia, which present peculiarities and deficiencies even yet more striking. Not one species found in the great islands westward inhabits Aru or New Guinea. With the exception only of pigs and bats, not a genus, not a family, not even an order of mammals is found in common. No Quadrumana, no Sciuridæ, no Carnivora, Rodentia, or Ungulata inhabit these depopulated forests. With the two exceptions above mentioned, all the mammalia are *Marsupials*; in the great western islands there is not a single marsupial! A kangaroo inhabits Aru (and several New Guinea), and this, with three or four species of *Cuscus*, two or three little rat-like marsupials, a wild pig and several bats, are all the mammalia I have been able either to obtain or hear of.

It is to the full development of such interesting details that the collector and the systematist contribute so largely. In this point of view the discovery of every new species is important, and their correct description and accurate identification absolutely necessary. The most obscure and minute species are for this purpose of equal value with the largest and most brilliant, and a correct knowledge of the distribution and variations of a beetle or a butterfly as important as those of the eagle or the elephant. It is to the elucidation of these apparent anomalies that the efforts of the philosophic naturalist are directed; and we think, that if this highest branch of our science were more frequently alluded to by writers on natural history, its connexion with geography and geology discussed, and the various interesting problems thence arising explained, the too prevalent idea-- that Natural History is at best but an amusement, a trivial and aimless pursuit, a useless accumulating of barren facts,--would give place to more correct views of a study, which presents problems as vast, as intricate, and as interesting as any to which the human mind can be directed, whose objects are as infinite as the stars of heaven and infinitely diversified, and whose field of research extends over the whole earth, not only as it now exists, but also during the countless changes it has undergone from the earliest geological epochs.

Let us now examine if the theories of modern naturalists will explain the phænomena of the Aru and New Guinea fauna. We know (with a degree of knowledge approaching to certainty) that at a comparatively recent geological period, not one single species of the present organic world was in existence; while all the *Vertebrata* now existing have had their origin still more recently. How do we account for the places where they came into existence? Why are not the same species found in the same climates all over the world? The general explanation given is, that as the ancient species became extinct, new ones were created in each country or district, adapted to the physical conditions of that district. Sir C. Lyell, who has written more fully, and with more ability, on this subject than most naturalists, adopts this view. He illustrates it by speculating on the vast physical changes that might be effected in North Africa by the upheaval of a chain of mountains in the Sahara. "Then," he says, "the animals and plants of Northern Africa would disappear, and the region would gradually become fitted for the reception of a population of species *perfectly dissimilar in their forms, habits, and organization.*" Now this theory implies, that we shall find a general similarity in the productions of countries which resemble each other in climate and general aspect, while there shall be a complete dissimilarity between those which are totally opposed in these respects. And if this is the general law which has determined the distribution of the existing organic world, there must be no exceptions, no striking contradictions. Now we have seen how totally the productions of New Guinea differ from those of the Western Islands of the Archipelago, say Borneo, as the type of the rest, and as almost exactly equal in area to New Guinea. This difference, it must be well remarked, is not one of species, but of genera, families, and whole orders. Yet it would be difficult to point out two countries more exactly resembling each other in climate and physical features. In neither is there any marked dry season, rain falling more or less all the year round; both are near the equator, both subject to the east and west monsoons, both everywhere covered with lofty forest; both have a great extent of flat, swampy coast and a mountainous interior; both are rich in Palms and Pandanaceæ. If, on the other hand, we compare Australia with New Guinea, we can scarcely find a stronger contrast than in their physical conditions: the one near the equator, the other near and beyond the tropics; the one enjoying perpetual moisture, the other with alternations of excessive drought; the one a vast ever-verdant forest, the other dry open woods, downs, or deserts. Yet the faunas of the two, though mostly distinct in species, are strikingly similar in character. Every family of birds (except *Menuridæ*) found in Australia also inhabits New Guinea, while all those striking deficiencies of the latter exist equally in the former. But a considerable proportion of the characteristic Australian *genera* are also found in New Guinea, and, when that country is better known, it is to be supposed that the number will be increased. In the Mammalia it is the same. Marsupials are almost the only quadrupeds in the one as in the other. If kangaroos are especially adapted to the dry plains and open woods of Australia, there must be some other reason for their introduction into the dense damp forests of New Guinea, and we can hardly imagine that the great variety of monkeys, of squirrels, of Insectivora, and of Felidæ, were created in Borneo because the country was adapted to them, and not one single species given to another country exactly similar, and at no great distance. If there is any reason in the hardness of the woods or the scarcity of wood-boring insects, why woodpeckers should be absent from Australia, there is none why they should not swarm in the forests of New Guinea as well as in those of Borneo and Malacca. We can hardly help concluding, therefore, that some other law has regulated the distribution of existing species than the physical conditions of the countries in which they are found, or we should not see countries the most opposite in character with similar productions, while others almost exactly alike as respects climate and general aspect, yet differ totally in their forms of organic life.

In a former Number of this periodical we endeavoured to show that the simple law, of every new creation being closely allied to some species already existing in the same country, would explain all these anomalies, if taken in conjunction with the changes of surface and the gradual extinction and introduction of species, which are facts proved by geology. At the period when New Guinea and North Australia were united, it is probable that their physical features and climate were more similar, and that a considerable proportion of the species inhabiting each portion of the country were found over the whole. After the separation took place, we can easily understand how the climate of both might be considerably modified, and this might perhaps lead to the extinction of certain species. During the period that has since elapsed, new species have been gradually introduced into each, but in each closely allied to the pre-existing species, many of which were at first common to the two countries. This process would evidently produce the present condition of the two faunas, in which there are many allied species,--few identical. The great well-marked groups absent from the one would necessarily be so from the other also, for however much they might be *adapted* to the country, the law of close affinity would not allow of their appearance, except by a long succession of steps occupying an immense geological interval. The species which at the time of separation were found only in one country, would, by the gradual introduction of species allied to them, give rise to groups peculiar to that country. This separation of New Guinea from Australia no doubt took place while Aru yet formed part of the former island. Its separation must have occurred at a very recent period, the number of species common to the two showing that scarcely any extinctions have since taken place, and probably as few introductions of new species.

If we now suppose the Aru Islands to remain undisturbed during a period equal to about one division of the Tertiary epoch of geologists, we have reason to believe that the change of species of Vertebrata will become complete, an entirely new race having gradually been introduced, but all more or less closely allied to those now existing. During the same period a new fauna will also have arisen in New Guinea, and then the two will present the same comparative features that North Australia and New Guinea do now. Let the process of gradual change still go on for another period regulated by the same laws. Some species will then have become extinct in the one country, and unreplaced, while in the other a numerous series of modified species may have been introduced. Then the faunas will come to differ not in species only, but in generic groups. There would be then the resemblance between them that there is between the West India Islands and Mexico. During another geological period, let us suppose Aru to be elevated, and become a mountainous country, and extended by alluvial plains, while New Guinea was depressed, reduced in area, and thus many of its species perhaps extinguished. New species might then be more rapidly introduced into the modified and enlarged country; some groups, which had been early extinct in the other, might thus become very rich in species, and then we should have an exact counterpart of what we see now in Madagascar, where the families and some of the genera are African, but where there are many extensive groups of species forming peculiar genera, or even families, but still with a general resemblance to African forms. In this manner, it is believed, we may account for the facts of the present distribution of animals, without supposing any changes but what we know have been constantly going on. It is quite unnecessary to suppose that new species have ever been created "perfectly dissimilar in forms, habits, and organization" from those which have preceded them; neither do "centres of creation," which have been advocated by some, appear either necessary or accordant with facts, unless we suppose a "centre" in every island and in every district which possesses a peculiar species.

It is evident that, for the complete elucidation of the present state of the fauna of each island and each country, we require a knowledge of its geological history, its elevations and subsidences, and all the changes it has undergone since it last rose above the ocean. This can very seldom be obtained; but a knowledge of the fauna and its relation to that of the neighbouring countries will often throw great light upon the geology, and enable us to trace out with tolerable certainty its past history. A consideration of the birds of Aru has led us at some length into this subject, both on account of the interest attached to it, and because we are not aware of any attempt to explain in detail how the existing distribution of species has arisen, or strictly to connect it with those changes of surface which all countries have undergone. The Birds and Mammalia only have been used for illustration, because they are much better known than any other groups. The Insects, however, of which I have made a very extensive collection, furnish exactly similar results, and were these, particularly the Coleoptera, well known, they would perhaps be preferable to any group for such an inquiry, from the great number of their genera and species, and the very limited range which many of them attain. In imperfectly explored countries, however, Birds are almost always better known than any other group, as a larger proportion of the whole number of species may be obtained in a limited time. I think it probable that I have collected more than half the birds inhabiting Aru, while I do not imagine I have obtained one fifth part of the Insects. The following is a brief summary of my collections in this class:--

Coleoptera	572 species.
Lepidoptera	229 "
Hymenoptera	214 "
Diptera	185 "
Hemiptera and Homoptera	130 "
Orthoptera and Neuroptera, &c	34 "
Making a total of	1364 species.

Lest the conchologists should think I have quite neglected their interests, I may mention, that I have collected all the land-shells I could find or procure from the natives. I have only obtained, however, 25 species. Almost all are Helices (20 species), some pretty and some of curious forms, but I am not sufficiently acquainted with shells to say how much novelty they present. It is remarkable that I have not found a single *Bulimus*, which in Celebes was the most abundant group; the few *Cyclostomata* are also small and obscure. Reptiles are scarce. I did not see a snake six times in as many months. There are, however, on the shores many sea-snakes, whose bite is very deadly. The natives spear and eat them. Lizards are rather plentiful in species and individuals; they are almost all plant-dwellers, and run on the leaves and twigs with great agility. The coasts swarm with fish in immense variety, and mollusca innumerable. A shell-collector would obtain a fine harvest, but I have been too fully occupied myself to attend to any of these last-mentioned groups; having often found the greatest difficulty in properly drying and securing my bird and insect collections in the rude houses, boats, and sheds I have been compelled to occupy. Damp, mites, ants, rats and dogs, are all enemies which must be guarded against with ever-watchful vigilance, and from all of them I have suffered more or less severely. Bird and animal skins require daily exposure to air and sun for weeks before they are dry enough to pack away. In this time they accumulate to such an extent, that it is a constant puzzle and difficulty to find places to put them in, so as to keep them free from ants, which establish colonies inside the skin, whence they sally out to gnaw the eyelids, the base of the bill and the feet; arsenic they laugh to scorn; and there is absolutely nothing that will keep them away but water-isolation, which again requires space and constant care to keep perfect. When to these are added insect specimens by thousands, requiring still greater care, the mere labour of watching the collections during the time they must remain exposed to the air, to sunshine, and often to artificial heat, is greater than a collector in a temperate climate, and residing in weather-tight roomy houses, can have any conception of. These remarks are merely my apology for not collecting *everything*, which stay-at-home naturalists often imagine may be as easily done anywhere else as in England.

Note Appearing in the Original Work

[1]The *Paradisea minor* was figured by Dr. J. E. Gray from life, with the breast-plumes displayed as above-described, in the 'Illustrations of Indian Zoology,' vol. i. pl. 37.

www.ingramcontent.com/pod-product-compliance
Lightning Source LLC
Chambersburg PA
CBHW021339290326
41933CB00038B/993

* 9 7 8 1 4 7 3 3 2 9 7 2 0 *